THE NATIONAL TRUST

Investigating

THE STORY OF

Farm Animals

By Gillian Osband   Illustrated by Peter Stevenson

Contents

The story of farm animals	2	Hunting and racing	18
'When sheep were king'	4	The poor man's pig	20
Improving sheep	6	The squeak and the whistle	22
The future of farming	8	Poultry	24
Oxen and donkeys	10	Goats	28
From ancient aurocks to the mighty White Park	12	Other farm animals	30
Food and farming	14	Rare breeds and the National Trust	32
Draught and transport	16	Answers	32

The story of farm animals

Farming has a very long history: as long ago as 4500 BC, people began to keep domestic animals. Thousands of years later, in the third and fourth centuries BC, Celtic settlers in Britain had developed agricultural skills. Archaeologists have unearthed Celtic ploughs and harnesses and even the sites of Celtic farms preserved in the chalk downlands and limestone hills of southern England. A good example is on the limestone spur near Corfe Castle in Dorset, where the remains of Celtic cattle enclosures and farm buildings are just visible.

Animals that used to roam in wild herds were gradually tamed, and people learned the value of keeping livestock. Although some were still killed for their meat and hide, many more were farmed to produce milk, wool, eggs and as draught animals to work on the land.

The Romans

Many of our farm animals were originally brought to this country by the Romans when they settled in Britain in the first century AD. The most useful animals were oxen, for ploughing the land, sheep and cattle for meat, milk and clothing, and the horse, of course, for powering their chariots. In fact, animals were so important that the elaborate Roman farmsteads and villas, like Chedworth in Gloucestershire, were built to accommodate animals as well as people.

CHEDWORTH, GLOUCESTERSHIRE

The feudal system

In 1086 William the Conqueror decided to commission a survey of the whole of England, so that he would know exactly how much land each of his tenants-in-chief owned, how they used it, and how much income he could expect from taxes. This information was gathered in the Domesday Book. The land in England was owned by the king, but he allowed tenants-in-chief to hold some of it in return for providing him with certain services and payments. They, in turn, employed others to look after their estates - this system was known as the feudal system. Near the bottom of the scale, a peasant, or villein, in Domesday England would work on the land and look after the animals. The records show that most villeins looked after two oxen, a cow, a pig, six sheep and seven acres of sown land for the lord, as well as keeping their own animals too, so they must have been hard workers. The Domesday Book tells us all sorts of useful facts about eleventh-century farm animals: sheep were the most common animals, but pigs were almost as popular. Goats, cows, oxen, horses, poultry and even bees were also kept as domestic animals.

The wool trade

Farming was *the* important industry of the Middle Ages. Sheep were paramount, and the best sheep farmers were monks of the Cistercian order. A typical Cistercian monastery is Fountains Abbey in Yorkshire (see p.4). Monks came to this isolated site from their previous home in York because they wanted to live very simply, devoting much of their time to prayer, so the lonely task of looking after sheep suited them well. Monasteries all over the country became rich from the sale of wool, and throughout the fourteenth and fifteenth centuries the wool and cloth trades were the basis of the economy in England.

Enclosures

Farming remained the largest industry in Britain, occupying about three quarters of the population right through to the mid eighteenth century. Landowners realised that they could farm more efficiently by enclosing their land with fences and hedges. Many agricultural workers lost their jobs as a result, and when common land around villages was enclosed, peasant families had nowhere to graze their animals. At the same time, cottage industries, such as spinning and weaving from home, began to die out as factories were built and expanded. Families often decided to leave the land, either to emigrate to the New World or to move to towns and cities.

Livestock improvers

To feed the growing number of people living in towns, more meat and dairy produce were required. Fields that were once needed for ploughing and sowing were now turned over to grazing. Livestock improvers like Robert Bakewell developed breeds to meet their needs: sheep with heavier fleeces, cows that produced more milk, or hens that laid more eggs. As fashions changed so many traditional, local breeds became unpopular - these are now known as **rare breeds**.

There is a section about rare breeds in every chapter, with a check list for spotting their names and features. These breeds are now living museum pieces, showing us the types of animals our ancestors farmed, but they are also vital for our future. Rare breeds are used as stock to develop new types of animals to suit changing fashions in farming. At Wimpole Home Farm in Cambridgeshire, Tatton Home Farm in Cheshire, and many other properties, the National Trust are actively helping to preserve rare breeds for this purpose.

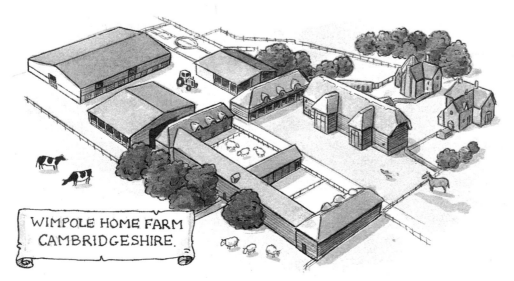

WIMPOLE HOME FARM CAMBRIDGESHIRE.

'When sheep were king'

Sheep have always been important farm animals, but in the Middle Ages and Tudor times, when the wool trade was at its height, they were a profitable investment.

The first sheep

There are wall paintings of sheep-like animals in the tombs of Ancient Egypt, but the history of domesticated sheep in this country really begins in the Bronze Age, around 2,000 BC. Archaeologists can tell a remarkable amount from the sheep bones they have found. For instance, Bronze Age sheep were smaller than our sheep today, and were farmed on higher ground. At Stannon Down on Bodmin Moor in Cornwall, the site of a Bronze Age settlement, remains of looms for weaving wool have been found.

When the Romans came to Britain, native sheep were small, horned, short-tailed and coarse-coated - similar to the rare Soay sheep you can see at Wimpole Home Farm in Cambridgeshire. Their wool was used for a rough cloth, but the Romans would not have been very impressed by its quality. They brought over white-faced, mainly hornless sheep with long tails, rather like the Cotswold breed, to cross with the native sheep and improve their fleeces.

Marshes and coasts

In Domesday England most sheep were farmed on rich pasture; fertile marsh land and coastal areas were perfect places for them to graze. The Domesday Book records large numbers of sheep on the marshes of East Anglia: some of these estates owned as many as 600 sheep (most tenants in Yorkshire and the Midlands owned under 100). East Anglia is still one of the main sheep farming areas in England.

FOUNTAINS ABBEY, YORKSHIRE.

Sheep rule OK

With the ever-increasing number of sheep being raised during the Middle Ages, there was more than enough wool for those who could afford it. Merchants from Flanders, Italy and France came to buy English wool. The monasteries, which were the greatest land owners and most successful sheep farmers, made vast sums of money from this trade.

Fountains Abbey

One of the richest monasteries in the thirteenth century was Fountains Abbey in Yorkshire. The Cistercian monks at Fountains kept their sheep on the extensive Abbey lands, and established outlying farms called granges. As the flocks multiplied they found that sheep farming took up too much of their time, so lay brothers joined the monastery to work on their granges. This meant that the monks had more time for their prayers and meditations. There is evidence that there used to be a wool house at Fountains where cloth would have been made, and we also know that the monks cleaned and dyed their own wool. Their rough, white wool habits probably came from their own sheep.

Granges

Granges were often some distance from the master abbey because sheep needed so much room to graze. Hawkshead Courthouse in Cumbria, for example, used to be part of a grange belonging to Furness Abbey, although it stands a good twenty miles away from the Abbey. Wool and other farm produce were gathered into enormous grange barns. These magnificent buildings reflected the wealth of their owners. The oldest surviving grange barn in Europe, dating from 1140, is in Coggeshall, Essex, originally part of the Cistercian monastery there.

THE FLEECE INN, BRETFORTON, HEREFORD & WORCESTER.

What's in a name?

Many place names in sheep-farming areas have their origins in the wool trade. The Cotswolds, a range of hills mainly in Gloucestershire and still the home of many sheep, was an important cloth-making area. The name probably originates from the word 'cote', a special building for sheep called a 'sheep-cote', and 'wold' meaning wild, hilly area where sheep grazed. Look at these other Cotswold names and notice the connection: Sheepscombe, Westcote, Kingscote, Condicote and Stow-on-the-Wold. Look out for pub names like The Golden Fleece or The Wool Pack too.

Wool families

Henry VIII put a stop to this great industry by destroying the monasteries. The King felt threatened by the power of the monasteries, whose lord and master was the Pope in Rome, not the King of England, and he needed money because he would indulge in expensive wars. Wool wealth passed to individual families - usually rich merchants or businessmen. One such family was the Sharingtons, who bought the convent at Lacock in Wiltshire and turned it into a house, now known as Lacock Abbey.

HAWKSHEAD COURTHOUSE, CUMBRIA.

A town built from wool

Lavenham, in Suffolk, was a prosperous wool trading town in the Middle Ages. You can still see the sturdy timber-framed houses built during that time. The crowning glory was the Hall of the Guild of Corpus Christi, built between 1520 and 1529 in the market place, when Lavenham was the fourteenth richest town in England. A thousand cloths were exported every year from the town, and the Guild controlled the sale of wool from the Hall.

Visit the Guildhall today to see a unique exhibition showing 700 years of the medieval cloth and wool trade. The trade mark of the wool merchants of Lavenham was a fleur-de-lys which you can see on the ceiling of the Guildhall, and in the plaster on some of the houses.

Improving sheep

HILL SHEEP, SWALEDALE

By the beginning of the eighteenth century there were four groups of sheep in Britain:

1) Small horned sheep with poor fleeces, found in Scotland and the Western Isles, little changed from prehistoric sheep.

2) Heathland sheep with coarser wool, found in the middle and south of England and Wales.

3) Black-faced, heavy-fleeced, horned sheep, found in the north of England, which were probably a cross between primitive sheep and medieval sheep imported from the Mediterranean.

4) Long-wooled sheep, direct descendants of Roman sheep.

LOWLAND SHEEP, WILTSHIRE HORN

CHARLES "TURNIP" TOWNSEND

Livestock improvers

Several livestock breeders, including Viscount Charles 'Turnip' Townsend from Norfolk, wanted to improve these sheep. Townsend grew turnips (hence his nickname) to feed animals during the winter. This extra feed had a great effect on new breeds of sheep because now they no longer had to survive on poor grazing land. It also meant that fewer animals had to be slaughtered in the autumn, and people's diets improved as fresh meat could now be bought during the winter months.

Robert Bakewell of Leicestershire was another great livestock 'improver' who developed new breeds of sheep by cross-breeding. Some of these new animals were rather an odd shape: his Leicester breed was known as 'barrels on four legs' because of their huge bodies, but

they produced excellent mutton earlier than other breeds. If you visit Townend in Cumbria you can see paintings of some very strange-looking sheep, bred by the Browne family who used to live there.

Thomas Coke, from Norfolk, was famous for his Holkham sheep and people came from all over England and Europe to watch the Holkham sheep-shearings.

King George III was nicknamed 'Farmer George' because he was fascinated by farming. He persuaded the Spanish to let him have some Merino sheep, renowned for their fine wool, and many of these sheep were sold at the first Royal Sheep Sale on 15 August 1804 in a field near London which is now Kew Gardens.

Did you know?

Robert Bakewell was the first breeder to hire out his rams. When he started his work in 1760, the rams were hired out at sixteen shillings (80p) a season. At the height of his success in the 1780s, he charged 2,000 guineas for the hire of seven rams - a vast amount of money at that time.

SOAY

MANX LOGHTAN

How local breeds developed

Breeds developed to suit their local environment. Hill sheep usually have long, heavy fleeces to keep out the cold, while lowland sheep have short, thick fleeces. Some have straight horns, others corkscrew horns and some are multi-horned, for example the Manx Loghtan sheep, which can have three pairs of horns!

HERDWICKS

Beatrix Potter and the story of Herdwick sheep

Without the efforts of Beatrix Potter, famous for her children's books, Herdwick sheep might not have survived. These sheep come from the Lake District, a beautiful area that has been threatened by development and tourism for 200 years. By grazing the upland slopes, Herdwicks kept the thistles and bushes under control, but during the 1900s they started to disappear from the Lake District because their rough wool was no longer in demand. Beatrix Potter was worried about the future of the Herdwicks - and the Lakeland countryside. With the money she made from her first book, *The Tale of Peter Rabbit*, she bought Hill Top Farm. She went on to buy fourteen upland farms, where she kept Herdwick sheep.

She became an expert on them, judging at competitions, and insisted that a flock of Herdwicks should remain on every farm after her death (now they are looked after by the National Trust).

Herdwicks are hardy sheep and some have been known to survive for up to a month under snow drifts by sucking the grease from their fleeces. They are now a rare breed, but the National Trust is working hard to save them. A carpet manufacturer in North Yorkshire has devised a way of making very hard-wearing carpet from Herdwick wool, which will help to keep interest in the breed alive.

The future of farming

There are over forty breeds of sheep in Britain, more than any other country in the world. Many, however, have become rare breeds, because of the changing fashions in farming. During the Second World War arable farming became very important as Britain needed to provide as much of its own food as possible. Pasture land which used to be grazed by sheep was turned into fields for crops, and this fashion continued after the war.

Rare breeds

Cotswold

These large, long-wooled sheep were brought over to England by the Romans, and have since adapted well to the limestone hills of the Cotswolds. The Woolsack, the famous seat of the Lord Chancellor in the House of Lords, is still stuffed with Cotswold wool!

Spotter's check list

Manx Loghtan	Black Welsh Mountain
Portland	Wensleydale
Herdwick	Wiltshire Horn
Soay	Jacob
Leicester Longwool	Hebridean
Lincoln Longwool	Shetland

PORTLAND

COTSWOLD

SHETLAND

Portland

The breed takes its name from the rocky Isle of Portland off the Dorset coast. There is a story that Portland sheep were brought over on the ships of the Spanish Armada in 1588 and swam ashore when the ships were wrecked. There is still a herd of Portland sheep in the 750 acre park at Calke Abbey in Derbyshire.

Shetland

Probably descended from sheep brought to Shetland by Viking settlers over a thousand years ago. They produce very fine wool, which is famous all over the world. The sheep are not shorn, they shed their fleeces.

Sheep products

Sheep provide a wide range of products: wool and milk while they are alive and meat, fat (tallow), skin for parchment, bone and horn after their deaths.

Roast saddle of mutton

Charles Francatelli, a famous chef who cooked for Queen Victoria, recommended roast saddle of mutton, cooked in front of an open fire. To make the meat taste especially good he advises the cook to 'dredge it all over with flour, shake a little salt over the surface, and with a large spoon pierced with holes, drop some dissolved butter over it, in order to give it that deliciously brown frothy aspect so well appreciated by all lovers of well-roasted English joint'.

Did you know?

Sir Harry Featherstonhaugh of Uppark in Sussex proposed to his dairy maid when he was seventy years old, and asked her to carve a slice out of the leg of mutton he was having for supper if her answer was yes.

Mad about mutton

In Georgian times mutton was a very common dish. Mutton pies used to be sold by a pie-man at street corners in most towns, and at the old country fairs in many villages (do you remember Simple Simon who met a pieman going to the Fair?). The pies were decorated with sprigs of mint to distinguish them from pork pies which were always topped with sage. The Victorians also enjoyed mutton. They usually had big families so large joints of meat were essential. Charles Dickens often mentions mutton in his famous Victorian novels: 'a boiled leg of mutton with the usual trimmings' was a common dish in most households.

Sheep search   Answers on page 32

Here are ten clues to help you find the words hidden in the box:

Member of the family in disgrace
A kind of knot
A female sheep
Aries sign of the zodiac
Embarrassed
Sheep in its second year
A kind of plant
Looking at someone in an amorous way
A sheep's coat
Hot drink of ale, apples, sugar and spices (1592)

```
Z L S X Y S T Q S P
G O H F D R A M E G
I O E S L A S E Y P
G W E U G E H S E E
E S P H A S E P S H
T B S C K B E C P Z
L M H C R M P I E F
B A A Q L A I S E L
P L N V V L S E H C
B R K I Z H H P S P
```

Oxen and donkeys

As far as we know, oxen were the first domesticated working animals, used for pulling ploughs and carts. They are remarkably strong beasts, hence the expression 'as strong as an ox' and, when yoked together in pairs could plough large stretches of land without tiring. Farmers used teams of as many as eight oxen, to plough the land in strips. This method of ploughing was designed to give everyone an equal share of the good and bad land: villagers each owned a strip of wheat or a strip of barley and looked after it themselves. It was not a very efficient way of farming, however, and was later replaced by enclosed fields.

Pictures of plough teams and other medieval farming scenes often illustrate medieval manuscripts, like the English Luttrell Psalter or the magnificent Burgundian book of hours, *Les Très Riches Heures du Duc de Berry.*

Worth a fortune

In the Middle Ages it cost about thirteen shillings to buy an ox, whereas a cow cost about nine shillings and a sheep only one shilling and five pence.

Did you know?

In 1957 oxen were still being used for ploughing on Earl Bathurst's estate in Gloucestershire. They were the last working oxen in Britain.

In Turkey, Greece, Portugal and most Third World countries, farmers still use oxen to plough the land as they cannot afford to buy modern machinery.

Mixed plough teams

The invention of the horse collar in medieval times enabled horses to be used as draught animals too (see p.16) and they were sometimes yoked together with oxen. An early eighteenth - century picture of a mixed plough team, with two single horses harnessed in front of two yoked oxen, was found in Eskdale, Cumbria.

Transportation

People relied on oxen to transport heavy goods in the Middle Ages: they pulled cart-loads of produce to market; building materials, such as lead, stone or sand from quarries to the towns, and they were particularly useful at harvest time for shifting crops into storage. By the eighteenth century, however, working oxen were being replaced by horses and later by tractors.

onkeys were also common draught animals, particularly in medieval times when they were used to pull carts and carry heavy loads on their backs. Merchants often transported their goods on donkeys, and usually travelled in convoy for safety. They followed packhorse routes, known as 'courseys', which were generally old Roman roads. These pathways were so narrow that two trains of donkeys or horses could not pass each other, and after long arguments one would have to step aside into the mud. Pedlars used mules to transport their wares. A cross between a male donkey and a female horse, the mule was considered to be an inferior animal, with a reputation for being stubborn.

Roasting on the spit

Whole oxen were cooked by turning the meat on a spit in front of an open kitchen fire. In the large kitchen at Cotehele, Cornwall, which dates back to the late Middle Ages, you can see where the roasting-spits were fitted for cooking different meats. The kitchen at Ham House in Surrey had one of the first mechanical spit-jacks (worked by a system of weights and pulleys), in operation in the 1670s.

The donkey wheel at Greys Court

Before the invention of hydraulic power, donkeys were the means of powering the water supply on many estates. At Greys Court in Oxfordshire, there is a Tudor wheelhouse with a donkey wheel inside. The wheel is mounted over a well, so when the donkey walked over the boards of the wheel, buckets of water were hoisted out of the water and up to a tank in the roof. A hook caught each bucket and tipped it into the tank, from where water could be channelled into the house.

Literary connections

In Shakespeare's play, *A Midsummer Night's Dream*, the mischievous fairy, Puck, plays a trick on a simple weaver called Bottom, giving him the head of an Ass!

There are many Biblical stories about donkeys, one of the most famous being the story of Jesus's journey into Jerusalem on a borrowed donkey. If you look at the marks on a donkey's back you will see a cross running over the shoulders and down the back.

From ancient aurochs to the mighty White Park

Farm cattle are descended from ancient wild beasts called aurochs, which used to live in the forests of Europe and Asia. Paintings of them still exist on the walls of Stone Age caves. Longhorn cattle look very similar, with wide horns and brown, mottled colouring, but the last wild auroch died in Poland in 1627.

When the Romans came to Britain the native cattle were mainly black with sharp, pointed horns. Only the old animals and some calves were killed for meat, because they were more useful as draught animals and for milk. The Romans improved the stock in Britain, and bones found in several archaeological sites show a wide range

in the size of their cattle. It is thought that the first white cattle were brought over by the Romans, although tradition has it that Celtic Druids sacrificed white cattle to their gods before Roman times.

Anglo Saxon and Viking settlers also brought their own cattle breeds to Britain and local types developed.

Cattle became increasingly important in the Middle Ages. Henry III granted a Charter to some of his nobles to enclose areas of forest as hunting parks in the thirteenth century. Herds of wild cattle, like the White Park cattle, were enclosed and hunted along with deer and wild boar.

AUROCH

DAIRY BREED. FRIESIAN

BEEF BREED. GALLOWAY.

Dutch influence

There was an increase in demand for dairy produce in Tudor and Stuart times, because of the growing population in towns. Farmers took advantage of the increase in trade with Europe and imported large numbers of cattle from the Continent. Many of our breeds today originate from Dutch imported cattle, famous for their high milk yield, like the Longhorn, Hereford and Gloucester cattle.

Improvements

In the early eighteenth century there was little difference in size and appearance between meat and milk producing cattle. Robert Bakewell, the master livestock improver from Leicestershire, started to develop breeds of cattle for particular needs. He developed hefty beef cattle, like the Longhorn, for the meat market, and dairy cattle to supply the growing population with more milk and dairy produce.

Today there are many individual beef and dairy breeds - as well as dual-purpose cattle. Dairy cattle like the well-known black and white Friesian, produce about twenty-two kilograms of milk per day. Jersey and Guernsey cattle produce thicker, creamy milk. Beef cattle have more meat on their hips and hind legs, thick necks and short legs. They look almost square in shape, like the Hereford breed.

Cheshire cheese

Cattle in Cheshire were crossed with Dutch breeds to produce greater quantities of milk. By the beginning of the eighteenth century Cheshire sent 14,000 tons of cheese per year to London.

BLICKLING HALL, NORFOLK

Prize cattle

Sir John Crewe, who lived at Calke Abbey, Derbyshire in the last century, was a keen cattle breeder. He came from a long line of livestock improvers and won several prizes for his Longhorn cattle. You can see the heads of some of these prize Longhorns on the walls of the Entrance Hall at Calke. If you would rather look at some living examples, there is a herd in the park at nearby Hardwick Hall.

LONGHORN

CALKE ABBEY

Bulls

Farmers and livestock breeders hired out their prize bulls for breeding purposes, and they often became worth large sums of money. Charles Collings, a cattle improver who developed the Shorthorn breed, sold one of his bulls, Comet 155, for 1,000 guineas in 1810.

A bull stands for wealth and virility, so it is a suitable symbol for a family crest. Sir Henry Hobart, who built Blickling Hall, Norfolk in the early seventeenth century, had a bull for his coat of arms, and two stone Hobart bulls guard the entrance to his home.

Bulls also have a reputation for being aggressive and some people believe that the colour red makes them extremely angry!

Cattle droves

Herds of cattle were driven along special roads known as droves to be sold at markets. In the seventeenth and eighteenth centuries drovers came from as far away as Wales and Scotland to Smithfield - the chief meat market for London. They shod their cattle before setting out on the great trek and would fatten the animals on rich pasture on the way. The Long Mynd is the main droveway from the north of England to Smithfield.

Food and farming

Cattle are generally farmed in rich lowland areas. With the introduction of chemical fertilizer, farmers can keep more cattle on their land because the grass is richer.

Preserving and preparing

Before the days of refrigerators most meat was preserved with salt, or smoked above a kitchen fire, then hung on hooks in a cool larder or meat house.

In wealthy households meat was eaten at almost every meal, with very few vegetables (only the poor ate vegetables so meat was a status symbol). On fast days and Fridays fish was eaten instead of meat. Hogarth, the eighteenth - century artist and caricaturist, made roast beef the epitome of patriotism with his famous painting, 'O, the Roast Beef of Old England'.

Superstitions and sayings

There are strange folk stories connected with dairy work. It is said that witches have special powers over milk, and may cast spells over a dairy. Milk that curdled or butter that would not set was believed to be the Devil's work. A silver sixpence thrown in the butter churn was supposed to be a protection from evil; so were three white hairs from a black cat's tail.

Dairies

From medieval times, self-sufficient households had their own dairies, which produced milk and other dairy products for the family and employees. Poor families owned a single cow or rented one from a local farmer, and returned it to him when 'dry'.

In the eighteenth century it became very fashionable for fine ladies to play at being dairy maids in beautifully designed buildings. While they wandered romantically amongst the cows or poured milk into pretty jugs, a real working dairy with a highly skilled dairy maid would be in operation nearby.

Shugborough dairy

An unusual dairy lies in the grounds of Shugborough Park, inside the Tower of the Winds. The dairy, on two levels, was converted from a banqueting room after the building was damaged by the Great Flood of 1795. You can watch butter and cheese being made at Shugborough Park Farm, exactly as it was in the eighteenth century.

SHUGBOROUGH, STAFFORDSHIRE.

Uppark dairy

The decorative dairy at Uppark, redesigned by Humphry Repton between 1810 and 1812, was a popular place for the gentry to pass the time, tasting cheeses and junkets, and watching butter being made. It was not the most practical place for dairy work, which went on in the scullery next door.

Rare breeds

White Park

WHITE PARK

These are one of the most ancient breeds of cattle, dating back to the first wild herds which were emparked in the thirteenth century. The most famous herd of White Parks can be seen in their original park at Chillingham in Northumberland, where they have lived for 700 years. These cattle are still completely wild, despite living inside a park, and will kill a calf if it has been touched by a man and has a human scent. The herd almost became extinct after the severe winter of 1947, when twenty cattle died. It took ten years for the herd to increase out of danger.

SHETLAND

DEXTER

GLOUCESTER

Gloucester

These are now very rare but were responsible for the original popularity of Double Gloucester cheese.

Shetland

This crofter's cow is capable of producing milk and good meat from poor pasture. Seven Shetland cattle were sent to the Falkland Isles after the 1983 war to help rebuild stocks as conditions there are similar to the Shetland Isles.

Dexter

A strain of Welsh Blacks found their way to Ireland in the Bronze Age; these developed into Kerry cattle. A Mr Dexter selected small Kerry cattle and produced the dwarf breed of cattle which was named after him.

Did you know?

There are also White Park cattle at Dinefwr Castle in Wales. During the Second World War this herd was used to plough the fields as there was not enough petrol for tractors, and the War Office insisted they were camouflaged in case the Luftwaffe's spotter planes used them to work out directions to Swansea and Milford Haven. The cattle were painted brown, but the change in colour had a strange effect - they started to fight each other!

Spotter's check list

Gloucester
Dexter
Shetland
Longhorn
Irish Moyled
Highland
Kerry
White Park
Red Poll
British White
Belted Galloway

Draught and transport

Celtic horses were small and wild - rather like today's Exmoor pony - until, once again, the Romans improved the stock. Their horses were bigger and stronger, used mainly for carrying messengers and travellers, and as war horses.

By the Middle Ages the horse had become a multi-purpose animal, used for ploughing, carrying knights into battle and as the most effective form of land transport until the invention of the combustion engine.

CALKE ABBEY, DERBYSHIRE.

Horse-drawn vehicles

By the seventeenth century horse-drawn vehicles were a common sight on the highways and in the towns. Road conditions were usually poor - especially in winter - so long journeys could be very uncomfortable. Wealthy gentry, who could afford their own horse-drawn carriages, travelled in great style and often at great speed. Most long-distance travellers, however, rode in stage - coaches. These heavy vehicles, pulled by four or six horses, ran regularly between the major towns. They stopped at 'stages' of about ten miles, usually inns, where fresh horses were kept for hire to the coach companies.

In the stable block at Arlington Court in Devon, you can see an impressive collection of over forty horse-drawn vehicles. There are also carriages collection at Calke Abbey, Derbyshire, and at The Argory, Northern Ireland.

Working horses

Horses began to replace oxen as draught animals in the Middle Ages. The invention of the padded harness in the fourteenth century enabled them to pull heavy weights and, before long it was possible for working horses to perform twice the work of oxen. Strong horses were bred to work as draught animals as well as for war. These breeds have gradually developed to suit local conditions: Suffolks from East Anglia, Clydesdales from Scotland, and Shires from the Shires of England (counties in the Midlands, such as Leicestershire and Northamptonshire). By 1800 almost every farmer owned a working horse. In the Lake District, where the terrain was often too steep for carts, horses pulled farm produce up the slopes on sledges, and horses are still used here to clear timber (a process known as 'snigging') because it is too dangerous to use tractors.

Ponies

Originally wild and hardy animals, ponies settled in isolated parts of Britain, such as Exmoor, Dartmoor, and the Shetland Isles, and local breeds developed. Ponies have always been useful pack animals, and in Victorian times pit ponies were used to carry heavy loads of coal, lead and iron ore in mining districts.

Shire

The largest and strongest of the heavy horses, these animals can weigh as much as 900 kilograms.

SHIRE

Suffolk Punch

Chestnut in colour, these animals live for many years (one stallion was recorded as travelling for twenty-five years without a break) and are renowned for their great strength. They used to do all the heavy work on East Anglian farms, and the Army employed them for dragging artillery into action.

Three Suffolk Punches are still kept in the Victorian stables at Wimpole Home Farm.

Dartmoor Ponies

When the tin mines were in production in Devon and Cornwall these ponies used to carry tin from the remote workings to the towns. Today the breed is in danger - the ponies are no longer as hardy due to inter-breeding with Shetland ponies - but a project is under way to breed 'true' Dartmoor ponies again.

Rare breeds

In 1939 almost every one of the half million farms in Britain had at least one working horse. But sophisticated farm machinery has threatened the future of the working horse. By 1970 a few hundred were still in use, but ten years later most of the heavy working horses were on the rare breeds list.

DARTMOOR PONY

SUFFOLK PUNCH

Spotter's check list

Exmoor Pony	Shire
Dartmoor Pony	Suffolk Punch
Shetland Pony	Clydesdale
Dales Pony	Cleveland Bay

Hunting and racing

'The sport of kings'

Horse racing became a popular sport in Britain in the sixteenth century. It has always been particularly enjoyed by the gentry and royalty: Charles II was obsessed with it and spent so much time at Newmarket in Suffolk that it became the seat of government for months of the year. (Newmarket is now the headquarters of British racing). His niece, Queen Anne, was equally devoted and in 1711 held the first race meeting at Ascot.

Towards the end of the seventeenth century, Godolphin Arabian, one of three Arab stallions, was brought over to England by Francis, 2nd Earl of Godolphin. **All** English thoroughbreds descend from these three animals.

Place your bets!

John Parker of Saltram in Devon used the money he was left when he inherited the family estates in 1768 to build a race course. Large sums of money were won and lost at Saltram Races - the Duchess of Devonshire, a hardened gambler, was a frequent visitor and is said to have found it an expensive day out!

There are also race courses at Blickling Hall in Norfolk, and Calke Abbey in Derbyshire.

Steeplechasing

A sport which started in the 1700s when country squires raced each other across the fields, from one church steeple to the next. Modern steeplechases take place on a race course, with fences and other obstacles to jump.

Hunting

Hunting wild boar and deer in private parks dates back to the Norman kings, and continued all through the Middle Ages. Elizabeth I enjoyed hunting the stag, and still went out regularly when she was an old woman (she died aged seventy). Fox hunting, made fashionable during the reign of James II, had originally been regarded not as a sport but as a necessity for farmers to preserve their livestock. It quickly became more popular than hunting deer, an increasingly endangered breed, kept mainly as attractive additions to country parks. Today fashions have changed again and drag-hunting, where horsemen and hounds follow a man-made trail, has replaced fox hunting in many places.

Stable blocks

Many National Trust houses have stable blocks, built years after the main house was completed. One of the oldest is at Dunster Castle in Somerset, which was probably built in the 1600s and repaired after the Civil War. When polo tournaments were held at Dunster at the beginning of this century the stables were filled to overflowing: the Maharaja of Jodhpur brought sixty-two ponies on one visit!

You can see another spectacular stable block at Willington in Bedfordshire, built in the sixteenth century by Sir John Gostwick, Cardinal Wolsey's Master of the Horse. It has an attractive gabled roof and high walls, and even includes living quarters and fireplaces for the grooms.

Other National Trust stables:

Kedleston, Derbyshire
Erddig, North Wales
Dunham Massey, Cheshire
Castle Drogo, Devon
Charlecote Park, Warwickshire

DUNSTER CASTLE, SOMERSET

The poor man's pig

Prehistoric swine roamed through the forests of Britain, living on a diet of beech nuts and acorns. In Domesday times the value of a forest - such as Hatfield Forest, once part of the Royal Forest of Essex - was measured by the number of hogs that lived there!

In the Middle Ages, poor families kept a pig in a sty beside the cottage, or at the bottom of the garden to fatten and kill (a second pig could be sold or kept for the next year). Pigs were the easiest and cheapest animals to keep as they could survive on left-overs. The family pig was killed in November to provide fresh meat until Christmas, and ham, bacon and lardie cakes for an entire year.

Did you know?

It was customary for Roman soldiers to swear an oath on a pig or piglet.

Sweet and sour porkers

Great changes in breeding began in the 1770s when pigs were first imported from China, brought in the holds of sailing-ships on the Far-Eastern run, to provide fresh meat for the sailors. The Chinese had experience of keeping pigs that went back 5,000 years, and had developed animals which fattened quickly. These pigs improved the British stock dramatically.

Regional diets

Local breeds were fed on the produce of the region. In cheese-making districts like the Vale of Berkeley in Gloucestershire, Gloucester Old Spot pigs were fed on whey, while Large Black pigs were fed on apples in the West Country or surplus cereal crops in East Anglia.

LARGE BLACK

GLOUCESTER OLD SPOT

Rare breeds

British Lop

One of the oldest and largest British breeds, these pigs come from the West Country where they were originally kept for eating rubbish! They have large lop-ears which flop over their eyes so they can't see very well.

MIDDLE WHITES

Middle White

These bristly, scavenging animals, dating back to the poor man's pig of the Middle Ages, were greatly improved when interbred with imported Chinese pigs (they have the same 'dished' faces). They don't like mud and prefer to live inside.

Tamworth

No one is really sure of the origin of this sandy-red pig. The most likely story is that Sir Francis Lawley of Tamworth was sent a wild red boar as a present from India, which he gave to an employee to breed with local sows in 1814, and the red colour was passed on. But some people believe that Sir Robert Peel created the breed when he imported a red boar from Barbados in the West Indies and crossed it with the local semi-wild pigs at around the same time.

TAMWORTH

Spotter's check list

Large Black	British Lop
Middle White	British Saddleback
Tamworth	Gloucester Old Spot
Berkshire	

21

The squeak and the whistle

The old saying that every part of the pig was used except for the squeak was very true. Black puddings were made from the blood, the soft entrails could be fried or made into pies called 'chitterlings', the trotters stewed with onions, the gristly ears, tail and snout made into soup or stews, the head, including the tongue made into brawn and the stomach stuffed and roasted for special occasions! The remaining meat was eaten as fresh pork joints or salted and smoked inside the chimney as bacons and ham. Even the fat was melted down to make lard, dripping and pork scratchings, the basis for warming puddings, pies and pastries.

The boar's head

*'The Boar's head
in hand I bring,
With garlands gay
in carrying.'*

In medieval times, when this ancient carol was written, wild boars roamed the countryside of Britain and were traditionally hunted just before Christmas. The animal's head was cooked and ceremoniously carried in to the Christmas Day feast, decorated with garlands of fresh herbs. This tradition still goes on at Queen's College, Oxford.

Candles

Until the beginning of this century, households had no electric lighting. Animal fat was supplied by the local slaughterhouse for making candles. If you didn't want to make them yourself (it was a smelly, messy job) the local chandler would provide tallow candles. Castle Coole in Co. Fermanagh, Northern Ireland has a tallow house (now the ticket office) which once made thousands of candles to light the servants' quarters.

In loving memory

The Countess of Mount Edgcumbe was deeply attached to a pet pig called Cupid. It followed her wherever she went, even on trips to London, and when Cupid died she erected a thirty-foot obelisk to his memory in the grounds of Mount Edgcumbe! It now stands on a hilltop overlooking Plymouth.

The Edgcumbe family also lived at Cotehele in Cornwall, now owned by the National Trust.

Make a piggy bank

You need:

a balloon
flour-and-water paste
pink water-based paint
newspaper
a cardboard egg box

Cover a balloon with small strips of newspaper, using flour and water paste to stick them (wallpaper paste is just as good). When one layer is dry cover the balloon with a second layer and wait until the glue has dried. Continue to build up about six layers until firm. Pop the balloon and pull it out through the hole where the knot was

tied. Cut out one piece of the egg box to make a snout and stick it over the hole with strong glue. Use four more pieces of egg box to make the legs and stick them to the bottom of the body. Cut a slot for coins at the top with a sharp knife (you might need some help here). Make ears for one end and a tail for the other. Paint the pig pink, and draw his face.

When the piggy bank is full you can tear it open ... or carefully shake the coins out of the slot if you want to keep your pig.

Poultry

The wild ancestor of all our domestic fowl is the Red Jungle Fowl which still lives in the jungles of India and South-East Asia. Once again, the Romans, with their taste for good food, brought several breeds with them to Britain (like the Dorking) and farmed hens and geese commercially for the first time. By Anglo-Saxon times poultry of all kinds was a regular source of food.

Hens

In the Middle Ages self-sufficient households kept a couple of hens to provide eggs and meat, while peasants kept fowl on the common land. Hens destined for market were locked up in fatting coops - like the one that Geoffrey Chaucer wrote about in *The Nun's Priest's Tale*, where Chanticleer the cockerel lived with his hens.

Cock fighting

Cockerels were not only valued for their time-keeping - some were bred for cock fighting. At West Wycombe Park in Buckinghamshire you can see a cockpit, an elaborately designed building where Sir Francis Dashwood and his friends watched this cruel sport before it was banned by an Act of Parliament in 1835.

New breeds

Eighteenth-century livestock improvers produced plumper birds and better layers, but the development of new breeds really took off in the nineteenth century, when poultry was imported from India and the Far East. The Americans developed some of their famous breeds at this time too - for example, the Rhode Island Red and the Plymouth Rock.

Weather vanes

A revolving weathercock positioned on the top of a high building or church spire shows the direction that the wind is coming from.

WEST WYCOMBE, BUCKS.

Henneries

Most modern-day hens live in enormous hen houses, and never go outside (these are battery hens). But before poultry was farmed on such a large scale, hens were left to strut around the farmyard, or a special poultry yard like the one at Wimpole which has individual timber boxes for each bird, and only shut in a hen house at night.

At Townend in Cumbria there is a most unusual hennery for hens and pigs! It is built on sloping ground so that the hens can use the top entrance while the pigs live underneath.

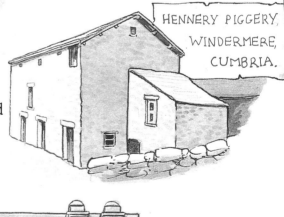

HENNERY PIGGERY, WINDERMERE, CUMBRIA.

Farming today

The huge poultry industry specialises in breeding birds either for their eggs or for meat. Egg breeds like the Leghorn or Minorca are usually small, active birds that start laying eggs when they are young, while meat breeds like Brahmas are larger, and don't lay many eggs. Free-range farming is becoming more popular as people are now more aware of the cruelty to battery hens. Look out for free-range eggs when you go shopping.

REDCAP

BRAHMA

BATTERY HEN

Rare breeds

Some breeds that are no longer fashionable, like the feathery-footed bantams, have managed to survive as ornamental show birds.

But many are now almost extinct in their pure form, and kept only for breeding purposes.

OLD ENGLISH GAME

Spotter's check list

Dorking	Scots Dumpie
Brahma	Redcap
Cochin	Orpington
Araucana	Surrey
Silkie	Old English Game

Scrambled birds quiz

These birds seem to have scrambled their eggs. Can you help them?

UCAARNAA =

OCHCNI =

YEOVCSMU =

NRAOM =

RFLKNOOKBALC =

OTCSSMUDPIE =

Answers on Page 32

Ducks and geese

There used to be many more geese on English farms: they provided meat, eggs, feathers, fat, quills for pens, and most important in medieval times, quills for arrows. They were a common sight on village greens and were driven in their hundreds to the London markets, particularly from the Fens in East Anglia.

Boarstall duck decoy

The duck decoy, near Aylesbury in Buckinghamshire, was built in the 1700s to catch ducks for food. You can see it in action today, but any ducks that are caught are ringed and numbered to study migration habits. The decoy man and his dog hide behind reed screens by the decoy pond. The dog runs in and out between the screens arousing the curiosity of the ducks who follow him. They are lured down a channel called a 'pipe' made of net-covered loops and caught at the end.

Doves and pigeons

In the Middle Ages these birds were a useful source of fresh meat and eggs during the winter, but keeping pigeons and doves was the privilege of the rich, so pigeon pie was a luxury rather than a poor man's meal. Young pigeons were known as squabs, and considered a great delicacy. Squab pie was made with the bird's feet sticking out of the pastry crust to identify the contents.

Dovecots

Wood pigeons were encouraged to roost in man-made pigeon houses from Roman times. By the Middle Ages pigeons and doves were housed in dovecots, with about 400 birds inside (although some dovecots had room for as many as 1,200 birds). They were often built in the gardens of great houses, like the medieval dovecot at Cotehele in Cornwall, because the droppings made good fertilizer for fruit and vegetables. It was also used to make medicine.

As manorial control relaxed, more people were allowed to build dovecots, and when the price of corn dropped in the late 1600s, dovecots appeared on many farms (a pigeon can eat its own weight in corn every day so you can imagine how much was needed to feed 400 birds!). They varied greatly in shape and size: the round dovecot at Kinwarton in Warwickshire contains 580 nesting holes in its walls, while octagonal dovecots were popular in the eighteenth century, like the one at Erddig, Clwyd.

Pigeon post

Their natural homing instinct, which brings them back to roost in the same place year after year, makes pigeons reliable carriers. During the Second World War these birds were used to carry secret messages over enemy lines.

Turkeys

NORFOLK BLACK

The first wild turkeys came from America in the 1520s, where they were kept by North American Indians. The Norfolk Black can be traced back to these first fowl. A turkey is the traditional bird for the Christmas meal and for American Thanksgiving - probably because it is big enough to feed the whole family ... and more!

Feathers

It was fashionable for ladies to decorate their hats and bonnets with feathers from the late seventeenth century onwards. Many breeding birds were killed to provide this decorative plumage, which greatly concerned a group of ladies from Didsbury near Manchester. They formed the Fur, Fin and Feather Group in 1889, which later became the Royal Society for the Protection of Birds.

Painted eggs

First you have to 'blow' the egg so that it does not go rotten. Make a hole at each end of the egg with a large needle. Place your mouth over the hole and blow gently, emptying the contents into a cup or basin. Carefully paint a design on to the shell with poster paints, or felt-tip pens. Try making bright patterns, floral designs, faces or funny fat people. They make colourful decorations for Easter.

Goats

ANGLO-NUBIAN

By the time the Romans landed in Britain, goats had been herded in Asia and parts of Europe for 7,000 years. Roman goats were good milk producers and improved our native stock (you can tell which goats are descendants of these Roman goats because they have fleshy toggles hanging under their chins). Until the 1600s large goat herds were kept for milk, cheese, and meat, and their hair was used to make rope and coarse hair cloth.

Rare breeds

Spotter's check list

Bagot	Pygmy
Golden Guernsey	Saanen
Toggenberg	Anglo-Nubian

GOLDEN GUERNSEY

The poor man's cow

Goats were kept by peasants who couldn't afford a cow you could buy a goat for two pence in the eleventh century, while a cow was worth about five shillings (25p). They were considered rather second-rate animals: the eighteenth-century livestock improvers didn't think 'the poor man's cow' was worth bothering with, but changes were made in the next century when foreign goats were imported, mainly from Switzerland, with high milk yields (such as the Toggenberg and Saanen goats). As a result, the native British breeds nearly died out.

Bagot

These goats are said to have been brought to this country by Richard the Lionheart. Returning to England through the Rhine Valley from his Crusade in the twelfth century he was given a herd of black and white, long-haired goats. They were put in one of his royal parks and kept there to be hunted. In the 1390s King Richard II had a good day's hunting with Sir John Bagot on his Blithfield Estate in Shropshire. The King was so pleased that he gave the royal herd of goats to Sir John, and they have been known as Bagot goats ever since. They even form part of the family crest.
This is the rarest breed of goats; there are now less than forty Bagot goats left in the world.

Golden Guernsey

These goats come from the Channel Island of Guernsey. They owe their survival to one person, Miss Miriam Milbourne, who hid many of the goats in caves on the island during the Second World War so that the occupying German army did not kill them all for food.

BAGOT

Farming today

Goat's milk is easier to digest than cow's milk, and can be made into a rich, strong-tasting cheese, or creamy yoghurt, so many farmers keep a few goats for extra milk. Some breeds are also kept for their wool, or for their skin. They will eat almost anything: plants, hedges, the bark of trees ... in fact, whatever they can sink their teeth into!

MOUNTAIN GOAT

Mountain goats

In the Highlands of Scotland, goats were as important and as numerous as sheep right into the eighteenth century. Goats are still kept in upland areas where the land is too poor or too steep for sheep or cattle. They can balance on narrow ledges and climb jagged rocks.

Leather

Fine leather goods, like gloves and purses, are made from goats' skin. Some of the highest quality leather comes from the soft skin of a kid.

COTEHELE

Fauns and satyrs

Bacchus, the Roman god of wine, had a group of helpers who were half-goat and half-human. There are illustrations of these fauns and satyrs in *The Bacchanals*, one of the beautiful tapestries which cover the walls of Cotehele in Cornwall. In the same scene you can see a boy Bacchus riding a goat.

Did you know?

The goat is the mascot for most Welsh army regiments.

Other farm animals

LOP EARED RABBIT

Rabbits

Rabbits were brought over from the Continent in the thirteenth century. In the Middle Ages peasants supplemented their diet by poaching them from the Lord's land, and only landowners were allowed to keep rabbit. They were considered a delicacy and often served at banquets as a special dish. Artificial warrens, called conygheres, were built in parkland for rabbits to breed in, and were enclosed by a fence or ditch to deter poachers. You can still see the site of the warren at Petworth in Sussex, called Cony Park.

Poaching!

William Shakespeare as a teenager is said to have been caught poaching deer from Charlecote Park, Warwickshire. He was brought before Sir Thomas Lucy in the hall of Charlecote, probably fined and possibly flogged. He then fled to London where he became a famous playwright. Shakespeare got his revenge by introducing Justice Shallow, a ridiculous character based on Sir Thomas, into his play, *The Merry Wives of Windsor*.

Deer

The Normans brought large numbers of these graceful animals to this country after 1066. They enjoyed eating venison so kept a supply of deer in parks, where they could be hunted.

During the twelfth century, fallow deer were brought over from the Continent, as they were much easier to keep in an enclosed area. Fences still had to be very high around deer parks, like the fence at Wimpole Hall, in Cambridgeshire.

By the fourteenth century there were over 3,000 deer parks in England, fifty in Wales and eighty in Scotland. The Lord of the Manor needed a licence to 'empark' from the king, who often gave him a gift of a deer from one of the royal forests as a reward.

SUDBURY HALL, DERBYSHIRE.

Many of the larger National Trust houses have deer parks on the estate. You'll see Jacob sheep grazing with herds of red and fallow deer at Charlecote, Warwickshire and the park at Dyrham in Avon has one of the oldest herds of fallow deer (Dyrham comes from the Saxon word 'Deor-ham' meaning deer enclosure).

Some deer parks contain elaborate buildings, designed to be eye-catchers rather than practical housing for the deer.

The eighteenth-century deer-house at Sudbury Hall in Derbyshire looks more like a fort, with corner towers and battlements, and also acted as a look-out for the hunters.

Dogs

Although dogs are not really farm animals, no farm would be complete without one. Breeds have developed for specific purposes, such as sheepdogs for herding sheep or retrievers for fetching game. Clumber spaniels were first bred by the Dukes of Newcastle at Clumber Park in Nottinghamshire. They are stocky, methodical dogs, well suited to flushing out game from thick undergrowth. Many large estates kept several packs of Clumbers and there is still one Clumber spaniel, called Basil, on the Clumber Estate, who is looked after by the wardens.

CLUMBER SPANIEL

SHEEPDOG

Dog spits

A small dog (usually a terrier) was sometimes used to turn the spit in front of the kitchen fire by running round inside a wheel. There is still a seventeenth- century dog spit at the George Inn at Lacock, Wiltshire.

Make a rare breeds farm

Trace a few animals from each rare breeds section on to thin card. Draw a flat base and side flaps so the animal can stand up as shown. Colour in each animal and cut them out to make your own rare breeds farm.

Use a large piece of cardboard and draw a plan of the farm and surrounding fields with a thick pen. Then make trees from twigs (stuck down with blu tac); bushes with cotton wool or moss; farm building, like a barn or a pig sty, from small cardboard boxes (even match boxes) and stand your animals all round the farm.

Rare breeds and the National Trust

Answers

Here is a list of just some of the places you can go to see rare breeds and learn more about them:

Wimpole Home Farm,
Arrington, Royston, Hertfordshire
Built by Sir John Soane as a model farm in 1794 for the 3rd Earl of Hardwicke to demonstrate the many new ways of growing crops and breeding animals. Today the Home Farm has become a centre for rare breeds, where you can see most of the animals mentioned in this book.

Shugborough Home Farm,
Milford, Staffordshire.
This Georgian farm is also a rare breeds centre, and contains a flour mill and dairy, restored to working order. Watch the flour being ground by the miller, then baked into bread in the kitchens, and butter and cheese being made in the dairy.

JACOB SHEEP

You can also see rare breeds at these properties:
Arlington Court, Devon
Jacob sheep, Shetland ponies
Calke Abbey, Derbyshire
Portland sheep
Kingston Lacy, Dorset
Red Devon cattle
Hardwick Hall, Derbyshire
Longhorn cattle, Woodland sheep
Hill Top Farm, Cumbria
Herdwick sheep
Charlecote Park, Warwickshire
Jacob sheep, red and fallow deer
Ardress House, Portadown,
Co. Armagh, N. Ireland
Irish Moyled cattle
Attingham, Shropshire
Jersey cows

ARDRESS HOUSE CO. ARMAGH

Tatton Home Farm,
Knutsford, Cheshire
The farm has been restored and demonstrates the role of a home farm in supplying food to a large estate in the 1930s. You can see farm animals, a working dairy, a corn mill, smithy, wheelwright's shop, and a deer park.

Cotswold Farm Park, Gloucestershire
Although not owned by the National Trust, The Cotswold Farm Park must not be missed. It was created in 1970 for the conservation of rare breeds and has the most varied collection in the country.

Page 9 Word search

```
Z L S X Y S T Q S P
G O H F D R A M E G
I O E S L A S E Y P
G W E U G E H S E E
E S P H A S E P S H
T B S C K B E C P Z
L M H C R M P I E F
B A A Q L A I S E L
P L N V V L S E H C
B R K I Z H H P S P
```

Page 25
Scrambled birds

Araucana	Cochin
Roman	Muscovey
Norfolk Black	Scots Dumpie

First published in 1991 by
National Trust Enterprises,
36 Queen Anne's Gate,
London SW1H 9AS

Registered Charity No. 205846
© The National Trust 1991

ISBN 0 7078 0134 6

Designed by Roger Warham,
Blade Communications,
Leamington Spa

Printed in England by
William Gibbons & Sons Ltd.